Beside Another Sea

Beside Another Sea

David J. Lythgoe

Copyright © 2024
David J. Lythgoe

The right of David J. Lythgoe to be identified as author of this Work has been asserted by him in accordance with the Copyright, Designs and Patents Act 1988

All rights reserved

No parts of this publication may be reproduced, stored in a retrieval system, or transmitted in any form or by any means, electronic, mechanical, photocopying, recording or otherwise without prior permission of the copyright owner.

British Library Cataloguing in Publication Data

A Record of this Publication is available from the British Library

ISBN 978-1-914199-78-3

Cover design © Alan Prosser

Cover photo: St. Columba's Bay, Iona

This edition published 2024 by *Renascentia*
Manchester, England

CONTENTS

FOREWORD BY REVD. DR ANDREW PRATT	ix
ACKNOWLEDGMENTS	105
BESIDE ANOTHER SEA	11
FIRST DATE	12
IONA. Waiting for the ferry	13
IONA SOLILOQUY	14
A SKYE MARRIAGE	15
ELEGY FOR A HERMIT	16
ACROSS THE POND	17
ALEXANDRIA	18
ASTEROID	19
DREAM DAY	21
BITTER LEMONS	22
CHARCOAL	23
CORONAVIRUS	24
LOCKDOWN SAFETY GLASS	25
A WRITER'S BLOCKDOWN	26
IN SCOTLAND NOW	27
SNAKE ON A FELL RACE	28
SCHOOL EXCHANGE	29
VAPOUR TRAILS	30
A BREAD POEM	31
A SONNET FOR POETS	32
INSOMNIAC POET	33
FREEDOM	34
SEVEN AGES OF HAIKU	35
DAYDREAM STREAMING	36
THE IDEAS OF MARCH	38
FIRST TIMES	39
GIRL IN A GARDEN	40

LEAVING INDIA BY RAIL	41
ENCOUNTERS WITH GANDHI	42
MANDELA DIED A WEEK AGO	43
A LANCASHIRE LANDSCAPE	44
SNOW	46
CAESAR IN COCCIUM	47
CUT GLASS BELL	49
TORBRECK, LOCHINVER	50
ON PENDLE HILL	51
DESERTED ESCALATOR	52
CONVEYOR BELT	53
CITY COMMUTER	54
ANONYMOUS	55
BOWLER HAT AND UMBRELLA	56
CLOUGH WILLIAMS ELLIS	57
NORMAN BAMBER	58
DAD	59
UNCLE FRED - A CENTENARY	60
SHOPPING WITH THE ICE MAIDEN	61
STILL LIFE WITH COFFEE POT	62
CAFÉ AT NIGHT	63
MEDITATION	64
CREATION RAP	65
DEAD BEAT	67
ARROWHEAD	68
STARDUST	69
THE TRAGEDY OF MEMORY	70
MAPS	71
TELL ME WHY	72
I CONFESS	73
LONG DISTANCE RUNNER	75
TV TORNADO	76

TOURISTS IN IRELAND	77
NURSERY BEDTIME	78
MOONLIFE	79
THE EXPERIMENT	81
ON FINDING A DEAD RAT	82
AFTER DARK	83
THE SANDS MEMORIAL AT THE NATIONAL ARBORETUM	85
THE MYSTERY OF BIRDS	86
CADENCES	***87***
THE LAST LAUGH	88
SHADOW BOX	89
A WORLD ABOVE	90
WAITING ROOM	91
COMING HOME	93
CHRISTMAS EVE	94
SOUTH DOWNS WAY	95
ST. MARTIN'S, WINDERMERE	96
DUST MOTES	97
LETTING GO	98
BLUE GRASS	99
YOUR LAST SPECTACLES	100
A POET'S MARRIAGE	101
TRUTH	102
ON PENDLE HILL	103
GRIEF	104

FOREWORD BY REVD. DR ANDREW PRATT

If you know David Lythgoe you will not be disappointed by this book of eighty-five poems. Simply read on. But if you're tasting his work for the first time, with no spoilers, let me give you an idea of what you're holding in your hand.

I have known David for many years. I have found him to be a quiet, self-effacing person. In consequence his writing in these poems offers an insight to a sensitive, reflective personality which, through this medium, has been able to give expression to emotion, to impressions, sometimes with humour, yet equally able to voice the grief born of love and loss.

David has won awards for his writing and our reward is this gift which I sense has been gestating for some time. His subject matter often quarries memory. He unearths feelings as much as facts. Often the description of a remembered scene becomes a metaphor for life's conundrums. His poetry frequently spans personal thoughts, yet he is never sentimental. Often the poems originate in a particular context – a holiday, a shared experience with his wife, lockdown and the unexpected sound of birds singing, the waves of the sea or some other pattern of nature. Much of the writing is observational, reminiscent sometimes of the poetry of R.S. Thomas. David is compassionate while the narrative of his verse occasionally twists offering a political slant.

His background enables him to merge a travelogue with classical literary, and scientific linguistic devices.

In short this collection is a selection of gems to hold, to treasure and to reflect upon. The cliché would be to say, 'enjoy'. This you will do, but expect also to be moved, perhaps to weep, to sigh, to laugh but, above all, to be grateful for such gifts, skill and insight that have been shared.

Thank you David.

BESIDE ANOTHER SEA

A pile of windblown leaves, knee deep
and dry, has gathered by my cottage
corner door. They are whispering amongst
themselves. They jostle with their angst,
unkempt and pitiful as boats of refugees
borne shoreward on a tide of war.

Last year, the leaves arrived as usual.
Haphazardly. While far away upon a foreign
shore, beneath a mathematical array
of parasols that stand like sentinels
the sun was browning limbs for those
who would not notice, care, or think about

bedraggled voyagers in search of peace.
Like falling leaves, in servitude to chance,
the refugees were swept ashore in drifts
for an incognito gardener to gather
from the surf, a salt-soaked bundle
wearing tiny shoes, with little legs
                              hanging loose.

FIRST DATE

Watching an infant guillemot, about to fledge
high on a barren cliff where every ledge
above a surging sea lacked living space,
I recalled the first time and the place
we met and how I never spoke; afraid
you'd think my silence masquerade.

Its parents, with their duty done had flown
and left their single chick to fly alone.
I watched it hesitate before I saw it dare
to try its wings upon the unaccustomed air.
I guessed it knew what wings were for and flew
as if it had new strength, and soon was lost to view.

Although not knowing how, I too began to fly
when we first met. I've often wondered why
I took so long to spread my wings and speak.
I should have known you'd be unique.
For fifty years we flew before your feathers failed
and left me on a narrow ledge in sorrow veiled.

IONA. Waiting for the ferry

A splash of children, suddenly unleashed,
descends like a clamour of seagulls to land
on a tide-wracked beach. This chattering wave,
released from classroom chalk and bleach
delights in spindrift pricking at its lips.

This is rapture testing a world of mystery
denied by time to those of us grown old.
One day they'll sail away to be force fed
on lies and adverts gilded with fool's gold.

They'll learn an alien language then
where loan sharks swim on city streets,
discovering how experience gained adds
one more link to memory's ball and chain.

Sea-urchins, starfish, sea birds' eggs and sand
are all they need to travel freely through
this pilgrim land. They need this freedom, graced
with unencumbered ecstasy, before
they join me with their innocence debased.

IONA SOLILOQUY

They say there are more galaxies than grains of sand.
                That's hard to understand
when on this beach I hold
                a quantum in my hand.
There are too many sea-worn grains
                to count, so I let them trickle
through my fingers answering
                the gravity that rules the universe.
And time. But does time matter
                when I search inside the space
beyond the limit of the sea
                where galaxies abound?
I watch a line of hollows riding gently home
                between wave crests,
reminding me of rest before they vanish
                on the shallow shore.
For I sit on this beach alone and looking back,
                I recall the hollow in the pillow
that you left,
                filled now with the eternal silence
of a voice I hear no more.

A SKYE MARRIAGE

The sun was half asleep and the peats still
waiting to be lit when I left my bed
to run a mile of rock-strewn rutted track.
There Blaven blocked the western sky beyond
a thatch of sheep-cropped grass that grew around
a narrow, secret bay. Not caring then that
pebbles bruised my feet, I stripped and plunged
where sea urchins and shells all shared their home
with sea anemones that stained with blood
the granite rocks. Then, surfacing, I saw
the blunted hills of Rum surrounded by
a sudden, golden wedding ring of cloud,
and in that fleeting, naked moment knew
that I was one with earth and sea and sky.

ELEGY FOR A HERMIT

For years the hermit watched
the sun loop over the hill, saw
moon-silver gather in the sea.
Flowers in crevices blossomed
and died. Lambs born on
the machair grew fat. He fell
in love with lightning, rain,
sea mist. Thunder. The island
was his cell. His cell an island.

His days took care of themselves.
Each owned its identity. Different
from all other days yet the same as
yesterday and the day before.
And all the days that had been and
were to come, until one day they
found the hermit cold and stiff.
They laid him down where
earth was deep enough to dig.
Brambles grow wild above him.
Bracken smothers his cell.
Suns loop over the hill.
Moon-silver gathers in the sea.

ACROSS THE POND

Hearing the geese of autumn leave they said,
while leaves were falling gently down,
"The nights are drawing in."
Spring blooms had fled,

and rain had put the summer heat to flight
before their patient woods began to write
on every tree that flamed in gold or red,
that sorcery of fire that gives delight.

Divided by a common tongue,
devising words that seem to us quite wrong,
without a wall across the pond they made us trade
our simple English for their language tailor-made.

It was no crime to introduce such cranks
as bankrolls, scumbags, sneakers, yanks,
xerox, zombies, sidewalks, rookies,
freeways, gasoline, hoods and cookies,

for though, in Autumn, we deplore them all,
they for an amnesty, conceived "The Fall".

ALEXANDRIA

The dowager wearing the faded pearls
is tired. She stands unmoved, her bony
ankles wrapped in the dust of centuries.
A scab of plaster falling from a failing wall
disturbs the prayers from sun struck minarets.
Does she envy her famous sisters basking
in the unrelenting heat far up the Nile?

A drone of muffled voices navigates
her narrow streets, climbs broken
balustrades and crumbling balconies.
Across the square, a roof-top line of washing
waves disconsolate. Its flapping message,
lacking words, comes tumbling down unread.

Infused with memories that swirl amongst
the battered taxis, cars and trams, here
history struggles to survive amongst the office
blocks. No majesty. No pyramids. No mystery
of sacred temples open to this bitter
lemon sky. Steel shoes and concrete
sandals clothe this lady's feet.

The quaking land that stole her elegance
destroyed far more than history. She
knows the world still trades behind her back,
but finds her peace beside a tranquil shore.
Remembering light that once directed
happy commerce to her busy quay,
she turns her face to Homer's wine dark sea.

ASTEROID
(After Wordsworth)

*Thoughts on hearing that a DART
(Double Asteroid Redirection Test)
has been launched to adjust the
orbit of the asteroid Dimorphos.*

Is then no nook of outer space secure
from rash assault by human footprints sown
on pristine purity? We know for sure
the hubris of mankind is overblown.
Could not our sacred sunlit moon endure
until we gifted it testosterone?
The universe and silence then impure,
we failed to recognise what we had done.
We could not leave alone the sanctity
of space that occupies the universe.
No! We must play our tricks on gravity
because our minds have always been perverse.
This voyage to disturb an asteroid
must make all saints and angels paranoid.

APPLE FALL

Looking up while watching
a monarch put to rest, I saw,
through a closed window,
an apple fall at just that moment.
Straight down. No mistake.
No sound at all. A silent call from
the gravity that Newton knew.
The umbilical stalk of a life
severed. Right on cue.
Clearly the perfect time
for a golden apple in its prime.
Its life fulfilled and therefore
ripe for leaving. Ready to join
the everlasting, random atoms
of the universe.
                No hurricane,
but somewhere across a distant
waste of seas, perhaps a host of
butterflies, rising together from their
feasting, fluttered their royal
wings, not knowing why.
                In empathy.

DREAM DAY

Today I wait impatiently. I dream
you dress for church, deciding what to wear
while putting on your smiling eyes that hide
the pain inside. You ask for help, and so

I close your necklace clasp and smell your hair.
Last night I heard you whispering,
but stretching out my hand felt only cold
sheets where you might have been awake.

Gathering dead leaves I hear you call that
coffee's made. Then in the shade where cherry
blossoms fall, I almost find you there, pond
mirrored in your favourite summer dress.

Tomorrow, I will walk a windswept beach
and stop to watch you stand precariously
on sea wet rocks in search of cowrie shells,
or hermit crabs and sea anemones

before the sanded footsteps that you left
are washed away and the salted safety
curtain of my dream dissolves in stinging
spray. Returning me to life.

BITTER LEMONS

Too long forgotten on the dusty shelf
lie desiccated lemons in a cut-glass
Tuscan bowl. They've lost the essence
of themselves to someone's indolence.
The zest they had has left them here
with shrivelled skins so iron-hard that
a sharpened knife will hardly wound.

These cysts of gilded sunshine grew
where he will walk no more, although
he still delights in barefoot dreams
of moss bound forest floors enriched
with lemon yellow stars of celandines.

Staring at the lemons now, he knows
his youthful appetite for life has dried.
He will not taste again the solar breeze
that licked their dimpled cheeks. Though
fallen from a different tree, he sees
himself reflected in their sad neglect.
How long for him before the blade?

CHARCOAL

He releases power locked inside a clutch
of charcoal sticks and with perspective
mutilates snow whiteness.
With depths of greys and blacks
he dimples oranges,
draws hollows fit for apple stalks.
He knows what magic is.

The lightness of the skull he found
on the moor surprised him.
The roughened texture of the bone
that colour would destroy cried out
for coal.
The more he worked, the more
the crooked shadows drew him in
until he fell into the depths behind the
the hollow bones.
He saw then what the eyes
of innocence had seen – an ecstasy
of sunbeams, fields and meadows.
New-born lambs on springs at play.
Not coal black wings above.
Not shadows gathering below.

CORONAVIRUS

Deserted streets and empty roads
are silent now except for symphonies
of song from birds that haven't heard
about the need for distancing.

They congregate around the feeders that I fill.
They wait impatiently to jump the queue,
then squabble hungrily. Their only care is food.
Blue Tits snatch a morsel, race for refuge,

wait a moment, then take their chance
with life or death by flying swiftly back.
Unseen beyond the garden fence,
the hunting sparrow hawk will choose.

LOCKDOWN SAFETY GLASS

Between two mirrored hands
unseen transparency keeps
faith with love untouchable.
Puckered lips that wish to meet
leave whispered imprints
where they brush the pane
before the visit ends.

At home a comfortable chair
with tinkling glass of ice-cold
anaesthesia will help
frustration drown.
No sound unless
the mirror speaks.

Here dust has settled
silently on every
polished plain. One finger traces
runes no one will ever read,
and indolence takes down
a scrap book seeking
solace in the opened
pages of the past.

Here, neither hedge, nor fence, nor
glass restrains, for there is
space that's safe in memories
unconfined by misted, breath defying
glass defying death.

A WRITER'S BLOCKDOWN

Creative teacher, please spare me some time,
I'm having some trouble with finding a rhyme.
The cat and the mat where it sat may be fine
but there must be a much better end to a line.

*Yes, you're right, for this Covid's a terrible mess
and I've closed down my class for our brave NHS.
Poets today need a theme far more suitable,
so reflect on the virus and be pharmaceutical.*

*Try writing some stanzas about a clinician
in search of a vaccine to help a physician.
Think for yourself, then stand up and be bold,
(though most poets' verses aren't worth any gold).*

But look, I'm in lockdown and life's a real bitch,
what's wrong with me wanting to make myself rich?
*Just listen to me and take heed of your teacher -
forget about writing and start reading Nietzsche.*

IN SCOTLAND NOW
(After Robert Browning)

Oh, to be in Scotland now that August's here,
and those who drive to Scotland from the far-off South
will share a dram, enjoy perhaps an Aberdeen steak
and then rejoice for old times sake.
While grouse face death from the shotgun's mouth
in Scotland now.

And after August, in September, when
the stags are rutting in the misty glen
little they know of those men who stalk,
where the grouse that fly in haste
from secret nests inside the heather
by lines of beaters have been displaced.
And rain falls thinly in Autumn weather
in Scotland now.

That's the sound of the guns you hear
the crackle of shot and the voices raised.
All will be well and they'll raise a cheer
in London hotels where those men are praised
who shot the grouse and the innocent deer
in Scotland now.

SNAKE ON A FELL RACE

Racing down the hill we suddenly
stopped. A snake was sleeping
unconcerned, as if nobody
ever came that way. Spirally coiled
she had gathered for herself
the sun-soaked comfort of a stone.

The diamond scales were proof
of what she was, but no one moved,
nor dared disturb her innocence.
This was a privilege so rarely seen.

On such a sun blessed day, no runner
hoping for a personal best would stop
to look. Nobody racing thinks to stop.

But maybe stops to think.
And so we did.
For in that moment, time
itself had stopped for us.

SCHOOL EXCHANGE

Pointing skyward, excitedly, our visitor
exclaimed - "Voilà, Behold - La France."
Far from home beyond La Manche
she saw one perfect cloud-made map
reminding her of home. Now here to learn
a foreign tongue, she used two languages,
forgetting in her joy how much she knew.
We stood together silently, watching
the Pyrenees dissolve into a liquid sky,
Normandy slowly fold into itself,
and the Bay of Biscay fill with foam.
I heard her sighing, almost crying –" Hélas."

Today, a mackerel sky released a flock
of sheep that hung around, as if waiting
uncertainly for a shepherd and his
jet stream dog to drive them home.
I wondered then if French skies also
know the languages of sheep and
speak to them on summer afternoons
as still as this. And would my pixie friend
recall her childhood memories and
looking up one day, she might exclaim with
great delight - "Voilà, Behold - L'Angleterre."

VAPOUR TRAILS
(For The Man Who Named Clouds)

Born too soon, Luke Howard never saw a vapour
trail quick-silvered straight, short lived
as hot breath dribbled on a frosted window
pane. No doubt he would have guessed
why haiku fronds burnt by the sun
on summer afternoons dissolve so soon.
                                        Unlike
their sisters' thundering epic poems
that decorate our winter skies for hours.

Prey of merciless spitting streams of jet,
feathery chameleons lasting hardly a minute
are fickle as the changing wind - outcasts
from the royal ranks of altocirrus,
nimbus, cumulus, cirrostratus.

If Howard with his head in the alto strati
had lived to see these washing lines of
frothy lace edged underwear laid out
upon a cloth of undiluted blue,
being one of the scientific literati,
he might have named them too. *Ephemerati*

A BREAD POEM

Poems, like proving dough will
sometimes rise surprising from the page.
Tied down by ink they are no more
than thoughts that skim like stones
across a pond until they sink.
No evidence then, except
the ripples that disturb.

But read aloud, mercurial words
will kiss the wind and drift
until they anchor in another
port and come to life.
Unrecognized at first,
some change their taste,
resembling wasps who've taken
off their paper gowns.
These fly from the tongue
on bitter wings, while others have
the grace of swallows
swooping on still air.
All poems, when given voice,
expand like proving bread
infused with the poet's breath.
They rise like a dream and fly.

A SONNET FOR POETS

Time lies before you, waiting every day
for you to wrestle with your themes without
procrastinated wishes or desires.
Time may be gone before you find a way
to rid your unforgiving days of doubt,
or contemplate on what your poem requires.
The gift of time is for you all to use
in writing down whatever truths you find;
not spending it in thoughts that introduce
dilemmas to your introspective mind.
So do not tarry lest the fleeting day
be done. You know that time will not be kind.
Too late, your verses all will fly away
like phantom dreams that leave you far behind.

INSOMNIAC POET

With a needle of dreams he embroiders
the velvet of night with strands of satin and silk.

Midnight melts the wax from his wings.
His tears drip into cold cocoa.

He scatters the leaves of the forest.
He ploughs through the snows of Olympus.

He prowls on the wastes of the tundra,
finds fragments of words under stones.

He sifts sand on the shores of Atlantis.
Then surfs on the crest of the rainbow.

When tangerine dawn lights the day,
he's assembled a spectrum from clay.

With his needle of dreams, he's patterned.
the velvet of night with satin and silk,

With silver and gold. With creating.
Fabricating with writing.

FREEDOM

When you're out on the fell
and you're not feeling well
and a cloud trails its skirts
and your breathing hurts
and you've bloodied your knee
in the crumbling scree
and the way ahead is hard to see
and you look around for a friendly face,
but all you can see is empty space
you must look to yourself to find relief
to numb the pain of your biting grief
for you're out on your own
and you're far from the town
where there's safety in spades
in the dead esplanades.
But you wouldn't change places
with one of those faces
that habits sustain,
that never felt rain
for you're most alive
when you have to strive
to gain a summit
for you're not a puppet
pulled by strings.
You've got Mercury's wings
attached to your heels
so you know how it feels
to fly like a bird
and have the last word
and that's FREEDOM.

SEVEN AGES OF HAIKU

Age of innocence
describes the infants in their
loving mothers' arms.

Experience marks their
early years when taking steps
they sometimes stumble.

Age of learning shows
that some may never win; their
vanity will fail.

The age of duty
brings responsibility
keeping faith with truth.

Next begins decline
when memory, body, mind and
heart begin to fail.

Forgetting faces,
names and places, confusion
and bewilderment.

Reverting now to
innocence, exhausted they
land in loving arms.

DAYDREAM STREAMING

This room is hot where sunbeams swoon.
The summer term will end this afternoon.
Here, Homer nods. But, soft! What light
through yonder window streams?
It is the bard of Avon calling us from
Birnam wood where Dolly Parton sings,
full bosom'd friend of each maturing son
that lusts for life; who having
seen the liquefaction of her clothes,
would taste of Flora's honey'd fruits.

Outside, the bees collecting pollen
dream of honey. But today
we have naming of parts.
This is the greasing box on which
you will find the nipple.
And here's the cocking piece.
Don't let me see
you sucking your thumb, just let this
drowsy numbness chain your senses.
Then slow to sleep, perchance to dream
of lambs at play and amorous shepherds
making hay.
They've left their flocks for party frocks
To dream of succulent lamb chops.
Twelve nights of rumpy-pumpy cakes and ale.

****

Straight off the page such thoughts
come bouncing, images and dreams
interpreting, or so it seems, a world
of loose connected streams.
But dead? They are not dead whose
words are read. The dictionary
tree still grows in poets' minds,
new leaves replacing those that fall.
So poet, fly from laptop dreaming
into space by trampolining.
Leave your vacant screen to die
and write yourself new poems bouncing.

THE IDEAS OF MARCH

Every year when Spring appears
I find I've lots of new ideas.
I then write down with dedication
all my newest inspiration.
But young or old, all seeking gold
there's one thing poets must be told.
Spring tides are higher than they think
and danger lies in pen and ink.
So, if they need to leave a mark,
they must beware the ideas of March.

FIRST TIMES

At seventeen, I faced the world head down
ignoring contour lines. I pedalled hard,
but thought I never moved as wave
on wave of hills flowed swiftly by
behind my clicking wheels.
                              Until
I had to rest in Langdale where
a man too old for hurry talked
of hills which I had only known
by name. How much I wish today
I could recall those first times when
for me on foot, the earth stood still
in Wasdale, Eskdale, Borrowdale.

Grown old, I envy those who've yet
to sit by Loughrigg Tarn or climb
Jack's Rake, whose futures lie in cloud,
or sun and rain-soaked, reddened cheeks.
How much I envy those I'll tell
one day about the glorious Jaws
of Borrowdale, and how I bathed
in pools I shared, one first delicious
time, with rainbow trout beneath
the double bridge at Grange.

GIRL IN A GARDEN

I have arrived at the caravanserai of
Samarkand where the scent of jasmine,
rising from the dust of centuries,
garnishes my camel's breath.
Sunlight is chasing shadows into corners
as the ebony of night concedes to light.

New-born and feather bright, electric
blue as an immaculate sky, a damsel
fly touches down on stainless steel.
Ignoring the coffee in my cup, she prefers
with her long tongue to taste metallic heat.

Beyond a wall of date palms, hungry
desert lizards wait. She could have stayed,
but when a sudden puff of wind disturbs
the pages of my book, with the silence
of a shooting star without a song
she flies away.

Oh, fickle girl whose life is quickly done,
was there too much, too much
for you to learn in one so brief a stay?

LEAVING INDIA BY RAIL

The Bilaspur station bricks begin to glow
as a hazy sun moves, ponderous and slow
across a yellow sky. Disconsolate now,
inside the railway yard a starving cow
is tearing up a cardboard box. Anyone
could play a tune on its xylophone ribs.
Its flicking tail disturbs the flies that feast
on blood. Soon they'll need another beast.

Outside the station yard a puff of air
gyrating gathers up a cloud
of rancid dust that pirouettes, tip-toed
before the breeze that sweeps the station road.
Dead tickets and the daily papers left
from yesterday, caught by the zephyr, drift
against a wall that ends their flight.
A comfort for the homeless through the night.

I climb the steps to find my train is late.
Nine hours with nothing to do but wait
and watch the tired sun dissolve
into an Indian night that will involve
me in a stifling, febrile, dusky heat.
Is this my parting gift so bittersweet,
where sun and air and people all move slow?
Though wiser than the indolence they show.

ENCOUNTERS WITH GANDHI

Dad's face was strange the day
he told me Gandhi had been killed.
I didn't understand, but everyone
became morose as if the world
had stopped.

Mr Banerjee came to live next door.
He walked the streets in his pyjamas.
Mrs Banerjee taught mum how to wear a sari.
She made little pies but forgot to put any meat in.

In Delhi, at Gandhi's house
I held your hand as we followed
the manufactured footprints
of his last one hundred steps.

In Agra, you sat where Diana sat.

When you returned to England,
you left a silent hollow thundering
on your pillow with the imprint
of your absence. Thirty years
later I hold your cold hand remembering how,
once upon a time, I thought the world
had stopped.

MANDELA DIED A WEEK AGO

Surely by now there's nothing left to say.
The noble man is dead and all is said
that can be said. But still the heads
of state, as if they fear the silence
that he leaves, unleash from reservoirs
of words their eulogies in spate
enough to drown the world in pools
of sorrowing love.

                        Respectfully,
they wonder what the water holds.

Too late now, they could have waved
their palm leaves high or decked his path
with cloth of gold. They've still to share
his enforced confrontation face to face
with God. Only through his eyes might
they see God's glory hovering with gentle
wings of peace.

           His freedom message,
(God be praised) not wrath, revenge or
soul-destroying hate, but simple suffering love.

A LANCASHIRE LANDSCAPE

Did you know the town where I used to live
where Saturday men raced scrawny dogs.
Where streets would ring to the sound of clogs,
and a thousand chimneys choked me with fogs?

Did you know the grey hill where I used to sit
watching open wagons shunt coal from the pit?
Beyond those sidings I knew far away,
more hills of grey would be growing each day.

Did you know the fields where I used to play
with winding gear that was always near
and the whistle that blew so that everyone knew
it was end of a shift - unless it meant fear!

Did you know the street where I used to walk
every day to the shop for a loaf of bread,
where a spectacled man looked kindly down
and ruffled the hair on a small boy's head?

Did you know the park where I used to ride
on a roundabout scraping my shoes in the mud,
or the path that led me into a wood
where the scream of a partridge froze my blood?

Those hills of my childhood are now all laid low
and it's hard to remember I once used to go
to the hills with a bucket to pick nuggets of coal
When mother was pregnant and dad on the dole.

They have taken and levelled my memories for good
where miners would hack through the shiny black gold.
No footprints are left where the old pithead stood
and widows in mourning could not be consoled.

All the wagons have left now that kept me from sleep,
though I hear their clanking still loud in my dreams,
for the hollowed-out earth has its secrets to keep
in the loveless cold arms of the empty dead seams.

Now immaculate streets will forever deny
those forgotten grey clinkers of cinders and waste
for the sweat ridden landscapes that once saw men die
by green fields and woodlands have all been replaced.

SNOW

Oh yes! I knew the coal town, waking early
to the sound of shunted waggons clanking
and everything black with dusty coal. Crossing
the main line on foot, I collected numbers,
painfully squirming, sitting on the fence
cutting into me, counting trains. Listening
to the drumming of every coming
juggernaut thundering down the gradient
into the town, shrouded and funnelled
with sulphurous smoke and a steam
laden hell through the length of the
cutting.

Last night, snowflakes came at me from all
directions out of a darkened sky. They died
on my windscreen in thousands. All night
the snow fell silently, gently and softly, feather
light by light feather, until this morning when
everything, living or not, was purified
by snow. Even the town is white today, but not
the town I knew before the developers came.

CAESAR IN COCCIUM

The Antonine Itinerary, that is to say, The Itinerary of the Emperor Antoninus was a register of stations and distances along the roads of the Roman Empire. In the Itinerary, Wigan is thought, with plausible evidence, to be the best fit for the Roman town of Coccium.

When Caesar came to Coccium, he camped on t'Market Square,
He didn't know as that wer't place they used fer Wigan Fair.
Now mayor o' Wigan heard o' this and thow't he'd get some rent.
"Owd lad," he said, "that's not a place weer thee con pitch thi tent.

We've geet some fellers comin soon wi roundabouts and swings
and carousels and dodgem cars and sich like other things,
so if tha wants fer stop reet here I'll have fer charge thee rent.
I've never seen the likes o' thee, tha's set a precedent."

But Caesar, he were quite a lad and so got quite upset.
Now look here mate," he sez in haste, "do'st think I'm gerrin wet,
Cos if tha does, tha'd best look out. Mi army's comin' next,
and if they have fer sleep outside I'll ave fer send a text."

With that he lifts his toga up and teks his smart phone out
exposing quite a bit o' leg. Then mayor, he gives a shout,
"Ay lad, tha's geet a touch o' gout, I really didn't ken,
I know just what tha's feelin lad, I've had the same missen."

"I'll tell thee what," sez Caesar then, "now tha's seen mi toe
I'll let thi have a recipe to pay for what I owe.
It's private like but mi sowjers swear you'll never come to harm
they eat it almost everyday and it works just like a charm."

"OK," sez Mayor, "it'd best be good, or I'll set mi colliers free,
they're very good at purrin an' they'll clog thi on yer knee."
Then Caesar gets his haversack and delves reet daewn inside
and fetches out a garlic clove and shows it mayor wi pride.

That's what tha needs," said Caesar then," it's ne'er bin known to fail."
"If what tha sez is true," sez mayor,"it sounds like holy grail,
but we've geet our own already -called Uncle Joe's mintballs.
Try some fer size and if tha likes, then tek some back fer't Gauls."

So Caesar sucked and sucked and smiled, an t'mayor were smiling too.
He didn't care for garlic but he knew just what to do.
"Ee lad," he said, "your garlic makes mi breath smell kinda funny.
Keep it for your sowjers cos we'd rather have yer money."

"Ah reckon we can do a deal since tha likes mi Uncle Joes
If tha buys a load for Mrs C., thi gout I'll diagnose."
"And what is more," sez mayor, "tha con stay on t' market square
But I don't want any trouble, after all, that's only fair."

So mayor and Caesar both shook hands and shared a bottle o' red,
browt all the way from Italy, at least so Caesar said.
When Caesar's troops turned up next day, as weary as could be
Mayor sold each one an Uncle Joe wi' a written guarantee.

It said if they if they don't satisfy, then bugger off back to Rome
Coccium's not a place for thee, tha's better off a' whom.
But then he had a change of heart, and much to their surprise
He offered to exchange each one for a pack of Wigan pies.

CUT GLASS BELL

Upturned, this bell becomes a chalice.
Refracted rainbows glisten where
hand carved facets interact with light.
Such glass is beautiful, but light alone
is not enough to shake it into life.
We know this bell will only speak
when someone takes it by the hand,
for ball and chain lie mute upturned.
Could we too be silent bells who hold
within our hearts unlimited wealth
of which we seldom speak?
Or are we where the Spirit dwells -
each one of us a sacred belfry filled
with mute, recumbent bells?

TORBRECK, LOCHINVER

Here stands a dwelling built of language from
the rugged, hardened landscape of the past.
Here ragged lichen, loose as sheep's wool caught
on barbs of wire wraps every tree with age.
Here ancient walls, cocooned in moss beneath
the silver birches keep a faithful watch.

Look closely though. This granite silence speaks
of hands that placed them stone on careful stone.
Delighting in solidity and strength
for years they've felt the pain of slingshot hail,
wind, sun and rain. Yet clothed in patience still
they stand. By rural roads from city streets
these stones have called you here where deer tread
soft on all the mossy dreams you ever dreamed.

ON PENDLE HILL

One hundredth of a second at f8,
you said, and handed me the
Japanese stopping-time machine.
No need to lock you in. When
you get home, some chemist's
skill will flush you into light. CLICK.

Now you'll stand, eternally impaled
against an English sky that's bruised
with heavy snow in Lancashire.
An alien pasted on an album page.
Reminder of the one who held
your camera gingerly and thought
of hanging strings of mulberry leaves
and kami-kaze silkworm moths.
.
Was snow on Fujiyama lying dormant
on the day you left? And did you
hear the spirits speak a Lanky twang
so different from your native tongue?
If one day you should turn the album
page in hope of finding admiration
in your children's eyes, will you recall
the chill of frozen snow on Pendle Hill?

DESERTED ESCALATOR

Operating alone, the empty machine
drones through the steady hum of nights
and days, outstrips each dawn as steps
are swallowed up inside the marbled
mall. Emerging thirty feet below
it climbs again from floor to empty floor.
Nobody moves. Nobody comes
through the stationary door.
Nobody's there.
These steps have never felt the weight
of feet for years. But still the steps climb
up the narrow hill. Perpetual motion not
complaining, waiting for someone, anyone
to come. Use me now, it seems to groan,
I'll still be circulating here alone when you
are dead and time itself will be forgot.

CONVEYOR BELT

We know them by their leathers, jeans
and slicked down hair, their rings and studs
and piercings. The paradox is: That in their teens
they are the same, yet in their childhoods

were so different. This other worldly brand
of poltergeists today would kick the world to
shreds if given half a chance, for contraband
rebellion grows unchecked inside their heads.

Somewhere along the belt, perhaps a bolt too long,
or non-conforming thread began the heresy
of not discriminating right from wrong.
Or are we blind in our hypocrisy,

believing that they have no right
to share our trough?
Who'll save them at their journey's
end when they fall off?

CITY COMMUTER

Self-satisfied, unseeing, he ignores
each workday morning glory out of doors.
His fellows, likewise keep their greetings short.
A curt 'hello,' or something of that sort.

Financial columns occupy his mind
in place of interaction with mankind.
His season ticket corner seat's reserved,
a just reward that, rightly, he's deserved

for year by year he's sung the bankers' psalms,
creating wealth to grease their fat cat palms.
When he retires the city will forget
this incognito, faceless silhouette.

ANONYMOUS

Anonymous I like to be.
for then nobody bothers me.
I like to use a *nom de plume*
before I walk into a room.

I'd like a handle to my name
but really don't want any fame.
A moniker would be OK,
or else a clever soubriquet.

I wouldn't mind a pseudonym.
It's better than an eponym,
the only clue to those who know
their cottage from their bungalow.

There's nothing like an epithet
to give a profiled silhouette.
A label or a tag I know
would be an appellation, though

innominate I'd like to stay
until I reach my dying day.
Then, when my coffin's left in earth
and no one knows my date of birth,

and bones and flesh return to dust,
and nothing's left except a crust,
remember me as one who died
but kept his private thoughts inside.

BOWLER HAT AND UMBRELLA

He was a serious man who lived alone.
They say he died while working in his cellar.
By and large he was not well known,
except for bowler hat and umbrella.

With neighbours he would seldom empathize,
his conversation limited to whether
the day would turn out fine or otherwise.
Talked all the time about the weather.

What defined him was the paperwork we found.
His baptismal record from an Aberdeen kirk,
a record of his marriage, tightly bound,
and faded pass into his place of work.

He'd kept his cash accounts for house and car.
Perhaps they were reminders of his spends
to match his hotel bills, both near and far.
They told us nothing of his range of friends.

All he left in black and white now lives inside
brown envelopes no use to anyone.
Only those who knew him when he qualified
might tell us if he had a sense of fun.

CLOUGH WILLIAMS ELLIS

Twice a day from Cardigan Bay the tide
walks past as if to wash the feet of trees
that would reclaim the wooded strand
on which this private vision stands.
Clough Williams Ellis heard the music
long before he wrote the score
that is this unexpectedly alluring,
perverse and improbable
Mediterranean Portmeirion.

Inspired by dereliction, here he pasted
town and village green with pastel
colonnades and soaring campaniles
that grew perverse above the shaded tiles
of porticoes amongst the rocks and trees.
                            His grand conception
harmonises boldness here with melodies
that sing their music from an operatic stage.
Could this be proof that maybe life has more
to teach than people know? This perfect score
developed from a thought that only a poet's
eye foresaw when built upon a lonely shore
still holding back the arboreal, invading
horde and cold encircling tide.

NORMAN BAMBER

Before the overture, he always trembled.
But when his baton rose he conjured
harmony distilled from silence.
This man, who gave me friendship and respect
and loved black peas at Wigan Fair expressed
his forthright views on names much better known
than his. He fired venom from the dark
caves of his eyes, his battleship chin destroying
heretics. And yet, with every last night concert
done, he had to ask - "Was that all right?"
He took me once to search for flying saucers,
believed in ghosts, investigated elemental
particles and space inside solidity.
He asked me once about the stars, the galaxies,
the binding force of energy, the vast immensity
of space. I told him all I knew, (which wasn't much)
but more than he could ever guess. His energy
was bound by harmony enriched by notes
and silences that touched his heart – and mine.
Too soon I fell in love. Then marrying left him.
"Like all the rest," he groaned as if betrayed.
He gave me freely all he could, and left
for me a love of music in his will. Then,
like his idol, Mozart, died too young, Aged forty-one.

DAD

Every Christmas on the old settee
I'd squeeze between my mum and dad.
The coal fire cast our shadows on the wall
behind. Dad sang ecstatic, shared
his husky voice with Handel on the
wireless.   But I imagined blazing
pictures when the coals collapsed.
Discovering a pile of vinyl 78s,
my finger pushed a wind-up gramophone
into speed. I made it squeal.

Beneaththespreadingchestnuttree
Thevillagesmithystands.
Thesmithamightymanishe
Withlargeandsinewyhands.

I no longer think it funny though.
Hearing again
I know that my Redeemer liveth,
I find more comfort in the silences
than in my father's gruff fortissimo.

UNCLE FRED - A CENTENARY
Died in France, September 27th 1918

Born too late, I never met my Uncle Fred
who died too soon and left my mother
and my aunts to grieve. Nor did he leave
a picture of himself behind. They say one

of the twins he fathered also died - of
pre-maturity brought on by learning
of his sad demise in Cambrai's fields
six weeks before the blessed armistice.

Two things I own of Fred who died a hundred
years ago – a book awarded for attending
Sunday School. Also a card, a precious token of
a brother's love one Christmas, for his

little sister Bess, my mum, in pencil signed,
"From Fred". Its simple message reads –
"The bells of Christmas sound o'er all the earth:
May they for you ring Happiness and Mirth."

SHOPPING WITH THE ICE MAIDEN

As cleanly sprung as a glacial mountain stream
and mirrored by transparency she licks
her Revlon lips.
With eyebrow pencilled cold
contempt, ignores the shuffling multitudes
that stumble by behind her back or burrow
through the guttered trash the length
of Oxford Street.

Beyond the plate glass screen her longing kneels
before a crystal throne of bright Chanel,
disputing eye-bright wishes with
desire for Gucci
waiting for her surplus gold.
She hangs unmoved,
perhaps too long, too hesitant,
impossibly restrained,
as if she's modelling for a helicoptered
bee that hovers above a flower's
sun warmed paradise.

Before she melts.

STILL LIFE WITH COFFEE POT
Van Gogh

Maybe not the first he did in Arles
but certainly a foretaste of his famous
canvas suns. Three lemons drunk
with sunshine lie untouched.
There is milk in the earthen jug
and the coffee pot is hot. Oranges
from Seville are waiting in the wings.

With whom does he expect to share
his early breakfast in this sudden
Spring? What anguished secrets
hide inside his crow-black pot
dispensing coffee on the morning
air?   The tablecloth, itself a dream
in blue, falls like a stream from
the forward edge. This is Niagara in paint
and yet impassive as the yellow wall
behind. No doubt
he'll take his friend outside to sit
where sunlight warms the kitchen
wall and share with him the kiss
of sunshine's fruit. He lifts the
gold-rimmed sunburnt cup to
take a sip, tastes darkness
rising from the blue-black glaze.
Devouring the day before
the streets begin to burn,
he leaves his still life's table still.

CAFÉ AT NIGHT
Van Gogh

Listen! Here darkness scrapes its iron feet
as diners come and go.
Hungry cats
beneath the tables dine on scraps
and a poet hungers for words.
Empty chairs are waiting to be filled.
Night beyond the circle of light
swims uselessly,
incapable of reaching in.

As impotent as time, stars watch in vain,
too far away to touch. They pursue
their cosmic plan while sparrows
root about for crumbs between
the poet's shoes.
These last few years the poet has drunk
deep from the ocean of grief.

Diners at the tables talk. Eat. Drink.
Then fade into the gloom.
Not one has noticed the poet,
or cared to throw him a line.
Not even a word.

MEDITATION

I have no envy left. I am content
to sit alone, knowing that the busy
world is somewhere else.
                       All I need is here,
within my arm's length's reach. The black
notes on the manuscript are silent. Waiting
while I concentrate on timing. How they
interact with complex phrasing.

Sometimes I forget the need to breathe.

Then I let my fingers touch the keys
and music like a hidden flight of birds
begins to fill the room.

That's when I start to live and will live on
until the music ends. As one day music must.

No one is here., No one but I will hear,
for this is my salvation. My inspiration.
My daily hour of meditation.

CREATION RAP

God was sleeping alone in his rocking chair
when he opened his eyes, saw that nothing was there
so he thought to himself what a waste of space
I ought to make a place for the human race.

This sort of thing just will not do,
I'll have to make a universe or two.
But the absence of light made it oh so hard,
though God was working in his own back yard.

So God clapped his hands with a thunderous bang
and light bounced around like a boomerang.
But light alone was not enough
so God made the air and some watery stuff.

He created the earth and it looked so great
that he thought it was time he might celebrate.
Instead, he added flowers and some woods and trees
some beasts for the fields and some fish for the seas.

He made a beautiful sun to shine by day
and he spangled the night with the milky way
Next he took some clay from a river bed,
made hands and feet and a body and a head.

Then he bent right down to give the kiss of life
and decided that the creature needed a wife.
Now Adam was the man so we believe
and the woman he created we call Eve.

The universe at last was all complete
so God took off his shoes and put up his feet.
He felt like a sleep so at last he lay down.
He pulled up the sheets and he took off his crown,

for everything he'd made he knew was the best
and settled himself down for a well-earned rest.
God knew that everything he'd made was good
But since he is our God, of course he would.

DEAD BEAT
(A Race Against Time)

Knowing I had a musical
ear but no metrical feet,
beat challenged me to a race.
I didn't like his tone knowing
it would be a matter of time
who won? He began
with a quickstep
and I was OK until
I heard the organ stop
to watch a foxtrot by.
It was a clef hanger
when we came to
the last waltz and
though I had time
to kill, I passed time
spending time
with Father Time
in the last bar.
Beating about the bush
and being out of breves
I quavered, but being a bar
behind the beat was drunk
and beaten. Dead beat.

ARROWHEAD

At first, we shaped flints
into axe and arrowhead.
We saw what axe and arrowhead could do.

Lightning gave us fire. We learned
to melt stones. We discovered iron. Fire and forge.
Forge and fire. Fire was light. Iron was power.

We saw an apple fall.
We shot arrowhead into the atom.
Everything was relative to everything.

We loved light, but we cancelled the stars.
Without light where could we go? We shot
arrowhead into space. But space was beyond.

Arrowhead spawned
a mushroom cloud from nothing. We
amazed ourselves at what we'd done.

Some day there will be nothing
left to know. And that will be the end
of Arrowhead and everything.

STARDUST

Stardust voices have chased me round
with whispering white noise. There's
no escaping tinnitus once it's found
a home inside your head. I lie in bed
wondering what it's talking about.
I'm not like the boffins at Jodrell Bank
who listen happily, because their voices
tell them how to keep their job and earn
an honest bob.

The voiceless refugees I hear are syllables
fizzing in an effervescent alien tongue,
a language built from nascent energy
transformed to sound from the intensity
of light released when gravity loses its grip.
Scientists listen because they wish to.
I listen because I must. Shut up stardust!
Shut up. **SHUT UP.**

THE TRAGEDY OF MEMORY

When the firing ended and the cordite
settled, the air was suddenly still. Fragments
of birdsong, left hanging on smoking trees
joined up to replace the screaming shells
with melody.

A million wondering, disbelieving men
climbed from the pockets of fear
in which they'd prayed. They stumbled
over groaning pockmarked earth
where pools of cold vermilion water
stained their boots. Unfamiliar freshness
blessed the air.

Delirious, they wandered off, as aimless
as the inconsequential whistling of delivery
boys riding bicycles on village streets. Woods,
flowers and fields for them were bandages
to dress their scars. But wounded minds
would not, nor could, forget.

MAPS

Before I sleep, I stretch my naked arm
across a cold unruffled sheet.
There's no response
so I let the radio fill the silent house.
That's what you miss. Any voice will do.
Thirties children, born into
approaching war, traced many paths.
Watched overlapping contrails left
by dog fights in the limpid summer air
dissolve. We knew what they were for.
Our future written on the sky.

That fading landscape of our elders,
became for us a signpost from the past.
Today, it's falsified by global warming
that's redrawn the map.
The faceless ghosts of lockdown left us
wandering in a maze of masks. Past, present
and future all collected in one message –
"Add to cart".

Across the hall, then up the stairs and counting,
that's where you live. At the centre of the world.
On the wrinkled backhands of the elderly
arterial lines of motorways pass slow through
brown-field sites where traffic flows until
the dynamo stops, and the map's withdrawn
and burned. Only the spider knows the way
through its own web. What did you expect
to find on a cold unruffled sheet.

TELL ME WHY

By all means tell me how, or what or who
or when. But mostly tell me why.

I know the earth revolves,
And flowers grow,

the seasons come and go.
But what I do not know

is why water falls from high to low.
Nor why in winter-time there's snow.

And why is every flake unique?
They surely would tell me why
If only they could speak.

And why is time unsearchable,
and why does love
equate to pain unspeakable?

Why must I wait until I die and have to grope
in hope, that one day God will tell me why?

And yet, as long as poets write, or breath sustains
this much I know, that love, in chains, remains.

What is eternity? If there's to be no end
when did it start? Please tell me why.

I CONFESS

I confess that I've been slow to understand
why a carpenter's son should wish to reconstruct
a tree when, from the first picture painted
by the first hand on the first wall of the first
cave we have never stopped painting.

I arrived when the elders spoke in tongues
veneered with the certainty of truth.
Simple prayers were then as flaking paint
on derelict walls. As irrelevant to youth
as the stony waste behind the reredos

where the high ground of the righteous
taunted us. Immutable as rock, they
slept in peaceful beds protected by
the holiness of sanctimonious sheets.

So we scratched ourselves a living space
and foraged among spiders' webs adorned
with insignificant deaths. When cornered by
the heat of doubt we realised that webs
were dangerous, though safer than
living face to face with scavengers
munching crumbs of garnered skin.

We grew bored and tired of reading lies
on scrunched-up paper scraps. The
clowns had shed hypocrisy and fled,
leaving us no clue on stone-cold shrines
to tell us what truth was. No light shone
forth from holy candles lying snuffed,
abandoned in the dust we dare not touch.

Roaming the earth, one day we came
across a field of buried words that spoke
a language tinged with burning fire and gold.
Putting our spades to work, flowers and trees
began to sing with words we almost understood,
and with simplicity we returned to painting, still
hoping and believing.

LONG DISTANCE RUNNER

Long distance runners know a secret kept
from the eyes of lazy readers of the script.
Such couch potatoes, fearing the race,
must blame themselves for failing in life's chase.
Your pilgrimage was more than one short length
of days in which you first developed strength,
when bursting lungs would not complain against
the happy rhythm that your legs dispensed.
Nor should you, looking back, regret those days
that will not come again, for you may gaze
today where roses never cease to grow.
How came you here? Was this by dreaming slow?
No! This is paradise, your race is done.
Elysium lies here beneath a golden sun.

TV TORNADO

Escaping   nosetotailing
nowhere parking
networklunching
texting sandwichsnatching
undergrounding   escalating
facelesscrowding   cominghoming
pouringgin.

****

Switching   onfor   watching
desert storming   metal twisting
towerscollapsing
unbelieving   Christchurchmurdering
innocence dying   Grenfell blazing
fireinvading   schoolkids running
earth plates sliding   tsunami
bursting   into living
rooms   drowningcoffee cups
upsettinggin.

****

Global warming   childrenwarning
climatechanging   daytimedarkening
lightning flashing   palmtreesbending
mainroadsmelting   forestsburning
Listening   watching   ending
refugecraving wantinghiding
round and rounding
noplacefindingpouringanother
anaesthetizinggin.

TOURISTS IN IRELAND

Sheep going to slaughter enjoy better deaths.
You will drink Irish coffee till you take your last breath.

You will sing Danny Boy, you will dance to their bands.
and flirt with the colleens on Kerry's gold sands.

You will sing till you're hoarse and your system rebels
and laugh at the jokes that your coach driver tells.

You will buy Irish cards as you drive round the coast
take hundreds of photos and clog up the post.

You will drink pints of Guinness, an old Irish brew
that's filled with the goodness their ancestors knew.

But you'll come back to England with memories galore
that will last you a lifetime - and probably more

NURSERY BEDTIME

Soft lies the head in a garland of roses.
Deep is the peace of the comforting voice.
This is where silence her bounty disposes.
This is where butterflies metamorphose.

Little eyes close to the music of stars.
Mum's story enters the shell of the soul.
Shadows hide secrets under the stairs
or settle in corners where mysteries prowl.

Shadowland lamplight hangs warm on the wall
where fairy tales flounce in a magical dream.
Here purity, virtue and innocence dwell
tucked into a promise of safety from harm.

MOONLIFE

Just think.
If we lived on the dark side of the moon
we'd never see a day-rise
because the moon axle
turns with perfect synchronicity
tuned to every six hundred
and seventy-two hours.

On the other hand, if we lived
in the eye of the man in the moon,
dawn would be known as **moonbreak**
and only happen as the earth rose
thirteen times a year
when every crater's rim
would become a zebra's skin.

There would be no shadows on
the moon's equator at **mid-moon**.
Only your feet encircled
in a pool of night.

The elderly, living at home
would be provided with **mooncare.**
Or if not, then sent to attend a **moon care**
centre.

Hospitals would provide **moonrooms**
for patients and visitors.

Children would go to **moonschools**
and apprentices be offered **moon release**.

Shift workers would look forward to the **moon shift**
And watch **moontime TV**
on their **moon** off.

Lovers would **moondream**
and go **daying** about

Criminals would commit **moonlight robbery**.
Probably carried out by earthlight.

But of course, all that's nothing but **moonshine.**

THE EXPERIMENT

The experiment began when the vacancy occurred.
Too many PhDs. on board at desks.

Nobody willing to swab the decks.
Recruited Bill. Drove HGVs. Raw boned. Tattoos.

Cucumber fingers. Trainers and jeans.
Do anything to realise his dreams.

Wore his white coat with pride.
Smiled every day. Wanted to learn.

Pulled rank in the Red Lion. Then
dropped a crucible. Smashed a burette.

Spilt sulphuric acid. God knows he tried.
But he went downhill slowly for six months.

Stopped smiling. Threw in the towel.
Returned to HGVs. Last heard, he'd gone

downhill too fast. On the M62. Heading
for Heckmondwike.

ON FINDING A DEAD RAT

Five days and nights of frost.
Snowdrops, hungry for life
have wakened to the light.
They hang their angel wings
like frosted tears, circling
in homage the profiled
body of a dead rat.

Perhaps I saw it once
rejected, hated, feared it.
Finding it now unexpectedly
alone I wonder, has it left
a family somewhere?
Was this a starving father, brother
loving mother too weak to search
for food? Why die here among
the snowdrops, surrounded
by the purest white of God's
enveloping love?

I know this crust of earth
is hard, but it's soft inside.
I'll fetch a spade.

AFTER DARK

Last night I lay awake in the dark, listening.
Wondering. Where are they now,
the lost lads of my early school whose
names I learned one fresh September
day in 1948?

Big John was first to leave. Inheriting
his father's fragile genes, he went
to join his dad too soon by
dying young.

And Spike, bespectacled and curly-haired,
who had a gift for music, would surely
have been a virtuoso pianist
had he lived.

Gordon, a comedian, became the friend
of all, but unexpectedly and sadly died
by accident when driving downhill
over unseen ice.

Gentle Roger, the poet, was the next
to go. My introspective friend who lived
enough, although his health was never good.

Unlike his close companion, fleet-footed
Derek, who from the wing scored many tries.
He didn't know me when I saw him last,
and died soon after, lingering with
dementia.

I much admired inventive Colin who
became a scientist. I heard he went
to live and work on esoteric particles
in Switzerland, but soon lost track.

Nothing do I know of all the rest whose
solitary paths diverged too far from mine,
except that yesterday, I learned
one more has died.

Tony was the friend with whom I shared
a double desk on our first day. Now saddened
by the news, I lie awake in the dark and wonder.
Where are the rest?

How many days will pass
before no one is left
to remember?

THE SANDS MEMORIAL AT THE NATIONAL ARBORETUM

Amongst the killing plots and military mounds
a hedged-in secret space retains a waste of tears.

Stillborn **A**nd **N**eonatal **D**eath**S** recorded here.

There are no sands in all the world that
hold such sorrow far from the voice
of everyday. The umbilical path
that leads you there is lined
with messages on painted
stones remembering life
soon done or unbegun,
displaying grief for
tiny rosebud lips
that never felt a
kiss. Each stone
could shatter glass.
But here, if they could
mend a broken heart, they would.

THE MYSTERY OF BIRDS

With the swirl of an opera coat, a million
starlings sweep daylight from the sky.
From where have these incredible
creatures come? Have they finished
with work and come here to play?

Or has God told each bird
the time and place to congregate?
These birds could be murmuring
God's words in hymns of praise
for us to sanctify and consecrate?

Quite suddenly, harmoniously,
the masque is done. Mysteriously
the message funnels into darkness.
Leaves us wondering. Was this
display the voice of God made visible,
for God speaks loudest not from light
but from the dark of Calvary?
These birds will resurrect at dawn
dispersing words like pilgrims. Each bird
an impudent prophet hoping to be heard.

CADENCES

THE LAST LAUGH

I'll never know what voice, off-stage, had
prompted me to move from my comfy
chair to sit beside you lovingly without
a word. Was that the moment when
I sensed the truth of which we never
spoke, accepting the inevitable fact
you'd always known was soon to be.

Laughter had kept you going
for so long, but now it seemed
there wasn't any laughter left,
as sitting side by side we allowed
a tide of tears to flow.

 Until a clash of spectacles
made us laugh and weep together.

Briefly,
I wondered if that was to be our
last laugh, yet hoping desperately
it wasn't so.

SHADOW BOX

The box on the wall doesn't know about shadows.
Its cold screen is waiting for a negative plate
to bring it to life. One touch of a clinical switch
will show electrons interacting with bones.
Or something less solid.

The sun doesn't know about shadows or when
they measured the line at twenty-three and a half degrees
North, as if to tell the sun "That's far enough."
Then named the line
Tropic of Cancer.

That moment when they found
the white shadow amongst the shades
of grey was when time stopped.
But the sun doesn't know.
And the box on the wall doesn't know.

But I watched your shadow shrink as the days passed,
until, as if the sun were directly overhead, one day
it disappeared. Even surgeons have no power
to halt the progress of the sun.

A WORLD ABOVE

He found the hill inside a tattered book.
Bewitched with romance, told his friend and took
her by the hand along a path that led
them both to paradise. The earth was hard.
Struggling to rise above a world that tried
to hide beneath a shroud of mist, a thin
sun failed to pierce the shadowed, silent air.
Nothing moved, although each footstep left
an imprint in the frost rimmed grass of fields
devoid of sheep. Seen from the summit cairn
and valley deep, a thousand pin-pricked lights
began to mirror stars as winter smoke rose
slow, untroubled, from a hundred hearths.
They drank of that chilled wine of happiness
before they had to turn, yet knowing they'd
go back, go back, go back. Until one day
in May he carried ashes in his pack.

The climb is harder now but he still goes.
The stream still chatters to itself each Spring
amongst the meadows, bright with new-born lambs
in which he knows she would delight. Each step
he takes she takes with him. And so he'll climb
the hill until the day he joins her there.

WAITING ROOM

They came to know the route by heart.
The M6 and M56. Three years
of watching road works change
their lives. The car park always full by nine.
The flagstone steps, the corridors, the stairs,
the nameless doors, the waiting and
the queues for signing in.

The haggard faces wearing paisley scarves
sat still and, mostly, didn't talk.
Young ones, more adventurously did.
Perhaps these were the ones
who thought they'd win.

Quite soon they began
to recognise the ones
they guessed would not return.
They knew their purpose anyway.

From time to time, new faces,
replacing those who'd gone before,
came wandering in - lost souls
in search of space amongst
the comfortable seats.

It wouldn't take them long to learn
the drill of numbered tickets,
the TV screens and waiting,
endless waiting, always waiting
to be called by
understanding nurses
shouting names.

They'd soon find their way
to the coffee shop,
the corrugated cups,
the plastic lids, the straws. Would
there be time enough to drink? Then,

the slow inequitable
emptying as every long day passed.
Until one day he left
a simple message on
an unresponsive answerphone.

COMING HOME

They met by chance beneath an ecstasy
of thundering rain. Some things you don't forget.
Shoots of love while wandering grew tall.
On sunlit paths where undiscovered hills
rubbed shoulders with the sky they climbed,
and wrote their dreams in waters falling from
the fells. They always left unsatisfied.

On frosty nights they had no fears when silvered
moons threw shadows on their path. They walked
with ease on mountains, crowned with midnight
tracks where diamond glory blazed.
Exploring stygian cracks that set their blood
on fire above the chasms of the dark,
they climbed. Defying fate, they never died.

But here, returning late, the sky dark
and the key turned in the lock, he hesitates.
He fears the silence she has left him with.
Inside the house the ticking seconds own
the room from floor to ceiling, wall to wall.
Despite the shadows living there, he knows
he'll excavate once more the wastelands of
their past, by time and distance sanctified.

CHRISTMAS EVE

I have kept the faith my love.
I've bought the tree, attached
the baubles, the angel and the lights.
I've garlanded it with swags of silver,
red and gold. A cataract of silver falls
like tears from every branch.

I've placed the Christmas bells you made
beside the crib you bought
in Bethlehem.
In case I might forget.

The radio sings carols as I bake mince pies
alone. I'm wondering why I do.

Card hangers overflow, the surplus
finding space on ceiling beams. Fewer now,
although my cards, and yours from last
year's list, have been despatched.

Everything my love, is as you would have wished.
Except for this - you are not here.

No mistletoe this year

SOUTH DOWNS WAY

The day we went to Beachy Head,
I guessed you'd find those ups
and downs too much.
The bus was best - and free. As
free as the space beyond the fence
that seemed a frail defence.

Five years later, I walk the South
Downs Way alone, remembering
how much that freebie meant to you.
I take on board a drink at Birling Gap
where passing faces seem to wear
a vacancy. Wistfully it seems, they look
towards the edge where sky and land
both meet the sea below. I suppose
these couples share a private joke.
                        The world, I think,
ends there, cut off from life
as suddenly as the moment when
I heard you wish for sleep and closed
your eyes for one last time.

ST. MARTIN'S, WINDERMERE

Why do I sit in a silent church taking refuge
from the noisy world outside these rough-hewn
slated walls, remembering the day we sheltered
here against a bitter wind and freezing rain?

Why do I try in vain to concentrate
on the great east window and the plaster
painted spandrels bearing texts that offer
solace to the sad and weary heart?

Why have I escaped from summer heat,
the tourist hordes and the traffics' roar
for this silent world that gives me time
to contemplate eternity? And why

do I now fill an envelope with notes, gift-aided
for a church I hardly know before I leave?
Perhaps because in here I've found the glue
I need to mend, as best I can, a broken heart.

DUST MOTES

That first time we met I knew
was when the arrow flew.
I watched you then with eyes so
different from the ones I use today
requiring spectacles. Sometimes,
I thought you never changed,
and yet your changing left me
mystified. Now only photographs
of moments that I snapped remain
to tell me what I thought
I'd never know of what we really are.
No more significant than dust motes
interacting with a sunbeam, swaying
in the fickle updraft of your final breath.

LETTING GO

Had he not loved so much he would have been
content to let each lightning hour pass by.

Had he not been content to let each lightning
hour pass by he would have counted every

precious day. Had he but thought to count
each precious day he would have known

there's no return. Had he but known there's
no return, he would have saved those moments

that they shared and so invested both their
futures in a still-to-be discovered land.

But love, tormented by the thought of no return
still lasts. So there are places where he hardly dares

to go and hills he dare not climb. And when he travels
to a distant coast where shells, sand, rocks

and hallowed waves embrace the shore,
he knows that shadows left there long ago
will fling again their stinging slingshot pain.

He finds it hard to walk those hills and sands,
though harder still to stay away. Such journeys now
he finds, are worth the price before each letting go

BLUE GRASS

What do I know of vacancy?
I, who viewed the world
from space on mountain tops
never understood your fear of lifts,
your preference for stairs,
your grief at finding
claustrophobia
on the sixteenth floor.

I've watched them change
the sheets and smooth the pillow.
Pull back the curtains.
Let in the light.
But I sit still. Behind
my eyes one memory fills
the space. Again and again
you told me I'd survive.

The place you occupied is
free to wander now from
room to room in search of space,
while here some atoms of the air
you breathed must still be mingling
silent with my own, as insubstantial
as the catalogue of dreams we never read.
So now, I let carbolic fill the vacancy
once occupied by Blue Grass from
your perfume spray.

YOUR LAST SPECTACLES

were in expensive frames.
Did glittering zirconia help
with the epiretinal membrane
macularopathy that made you
hesitate on roundabouts?
Trying to cope with near sight
I wear them now to type.
You lost them briefly once.
So why, soon after, did you
have to leave them behind?
These are the ones I keep.
The rest all went to Ghana.
Yours give a clearer view
than my varifocals. Which
surprises me. You must
have seen the faces of
our friends as clear as
crystal, the furrowed skin,
the colours of their eyes.
I realise now why you
cleaned them every day.
When you looked into
the surgeon's eyes did
they help you read behind
the untruth of his smile?
Yet you missed the killing
bathroom step. I try
to see the world you saw.
But think I never will.
Perhaps I never did.

A POET'S MARRIAGE

My love
    I wish that you had learned
    how I, rehearsing every word
    was seeking perfect harmony.

My love
    you did not understand that
    poems are not planned but have
    their own lives dripping from the pen.

My love
    I chose my words like flowers.
    Shaped garlands for your love. Although
    most were unfinished, some had merit.

My love
    I miss your disapproving look,
    your put me downs, your finding
    fault. Your eagle-eyed examination.

My love
    I miss your condescending stare
    your disappointed voice suggesting that
    my choice of words was wrong.

My love
    a lightning bolt one day will strike the
    strongest oak destroying life and all
    that might have been. But most of all

My love
    I miss your gentle touch, the comfort
    from the mingling of our tears, repentance
    and the mutual forgiveness in our eyes.

TRUTH

Writing now in a minor key, I've learned that one day
two-faced love will turn
its back on you. No one had ever told me how love hurts.
That joy is just a joke.
You went away before the outer circles of your ripples
disappeared.
Before a cold, indifferent world would try to silence me.
When I choose the feather

duster that you held for cleaning, I hear the beating
of the metronome
that kept you breathing week by febrile, torrid week
until you fell
in search of the invisible dust I thought was never there.
And so it seemed
that death, once far behind, had caught me up to teach
me cruel truth.

Disconsolate then as if I stood behind a mirror
on the wrong side of the glass,
I heard myself reflecting how I held infinity in my
hands and saw my image changed,
replaced by all that's beautiful. That was the moment
when I realised
that I would never learn the answer to that one eternal
question: "What is truth?"

ON PENDLE HILL

Where are you now my love?
Still treading the paths where
we left you on the cold plateau,
free as the air and shaking your hair
to the wind's delight? I sing you my
songs though deep in earth I know
you cannot hear. The same earth
that felt the touch of your feet,
your hands, your labour, your laughter,
your tears.

Today my words are mangled by the alien
fires of a speech you'll never hear
as my towel rail collapses and I swear
at the toilet seat that needs screwing down
and the reading lamp that's failed again.
But I will forever return again
and again and again to the hill.

GRIEF

Grief is wanting the unattainable.
    Learn to live without.

Grief is having a long memory.
    Do not be afraid.

Grief is loving another too much.
    Take care of yourself.

Grief is looking too far ahead.
    Live for today.

Grief is knowing there will be first times.
    Get them over with.

Grief is knowing exactly what he/she would say.
    Be glad and listen every day.

Grief will be your shadow to the end of time.
    Do not look back *too* much.

Grief is an eternal expression of what love is.
    Thank God for the love you knew.

ACKNOWLEDGMENTS

I wish to thank the following for their assistance:
   Revd. Dr. Andrew Pratt for the preface.
   Alan Prosser for the cover design.

ALSO BY DAVID LYTHGOE

A House Nigh Unto Heaven (1994)
Travelling Light (2000)
Wigan Bred (2005)
Myrrh From The Forest (2015)
A Distillation of Hills (2022)

ABOUT THE AUTHOR

David Lythgoe was born in Wigan two years before the start of World War II and wrote his first poem (now thankfully lost) while still at junior school. During his career as an industrial chemist, he worked in India where, in 1988, he began to write seriously. This is his fourth collection of poems, inspired by his world-wide travels, the natural world, our place in the universe, and a (still) developing love of being alive. He was widowed in 2014 and is proud to have two happily married daughters. His poems have appeared in *The Daily Telegraph* and *The Methodist Recorder*.

Anrdrew Motion, Poet Laureate emeritus, wrote about one of David's poems "I like the way it evokes the past ... and yet connects with the present ... at the same time containing an element of lightness – even of joking – which is entertaining to the reader."

Revd. Dr Andrew Pratt writes: "David has won awards for his writing and ... he unearths feelings as much as facts ... His background enables him to merge a travelogue with classical literary, and scientific linguistic devices. In short this collection is a selection of gems to hold, to treasure and to reflect upon. The cliché would be to say, 'enjoy'. This you will do, but expect also to be moved, perhaps to weep, to sigh, to laugh but, above all, to be grateful for such gifts, skill and insight that have been shared."

Milton Keynes UK
Ingram Content Group UK Ltd.
UKHW020953241024
449966UK00009B/148